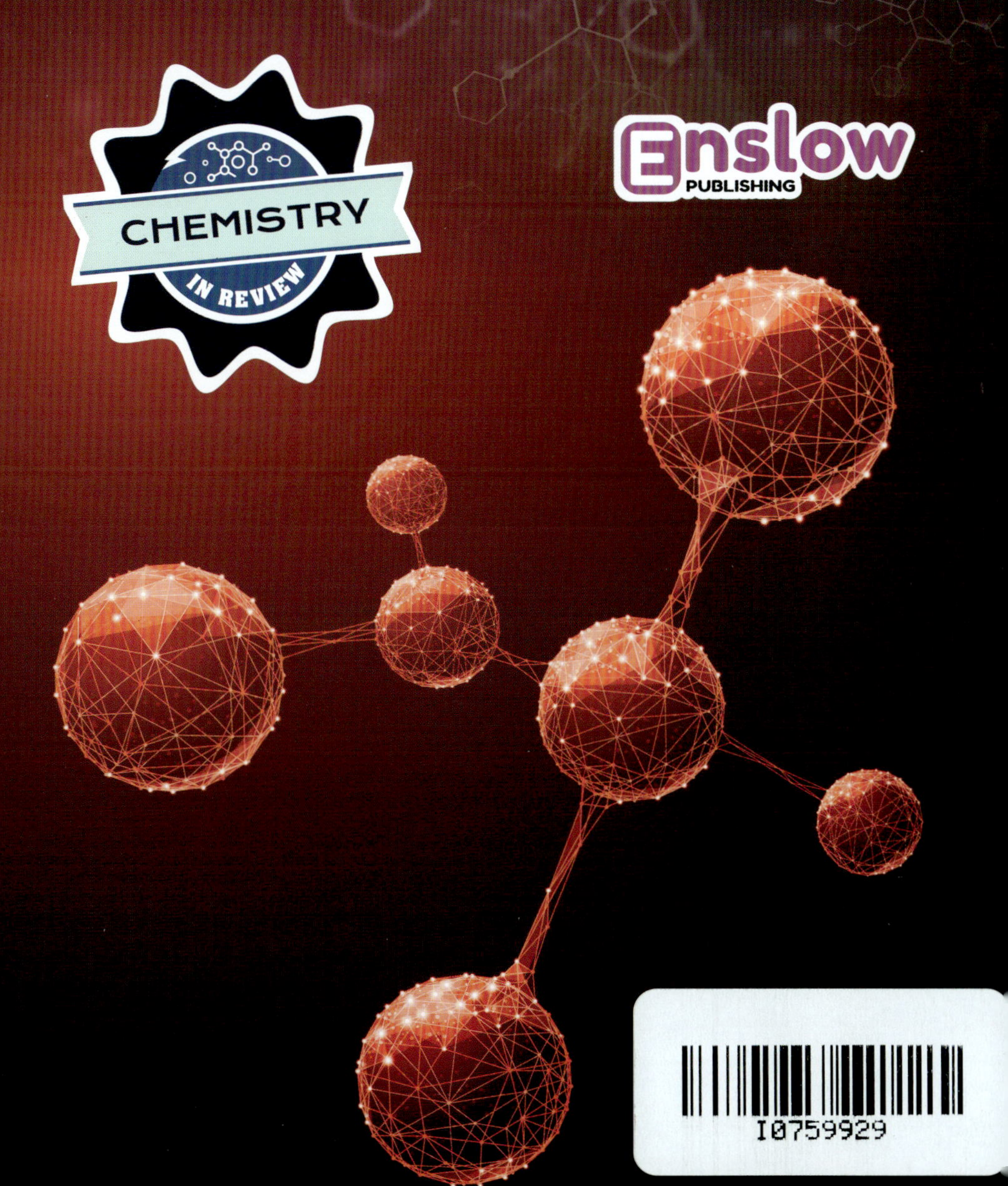

MOLECULES

BY MARIE ROESSER

Please visit our website, www.enslow.com.
For a free color catalog of all our high-quality books, call toll free 1-800-398-2504 or fax 1-877-980-4454.

Library of Congress Cataloging-in-Publication Data

Names: Roesser, Marie, author.
Title: Molecules / Marie Roesser.
Description: New York : Enslow Publishing, [2026] | Series: Chemistry in review | Includes bibliographical references and index.
Identifiers: LCCN 2024037090 (print) | LCCN 2024037091 (ebook) | ISBN 9781978542914 (library binding) | ISBN 9781978542907 (paperback) | ISBN 9781978542921 (ebook)
Subjects: LCSH: Molecules–Juvenile literature. | Chemical elements–Juvenile literature. | Chemicals–Juvenile literature.
Classification: LCC QC173.16 .R6475 2026 (print) | LCC QC173.16 (ebook) | DDC 541/.22–dc23/eng/20241112
LC record available at https://lccn.loc.gov/2024037090
LC ebook record available at https://lccn.loc.gov/2024037091

Published in 2026 by
Enslow Publishing
2544 Clinton Street
Buffalo, NY 14224

Portions of this work were originally authored by Kennon O'Mara and published as *Molecules* (A Look at Chemistry). All new material in this edition is authored by Marie Roesser.

Designer: Claire Zimmermann
Editor: Therese Shea

Photo credits: Cover, p. 1 (main image) AntonKhrupinArt/Shutterstock.com; series art (molecule header image) jijomathaidesigners/Shutterstock.com; p. 5 Inside Creative House/Shutterstock.com; p. 7 Bjoern Wylezich/Shutterstock.com; pp. 9, 11, 30 (molecules) LDarin/Shutterstock.com; p. 21 (top) Piyaset/Shutterstock.com; p. 23 (top) LuckyStep/Shutterstock.com; p. 25 (top) Zerbor/Shutterstock.com; p. 25 (bottom) Ingrid Liem/Shutterstock.com; p. 27 DUSAN ZIDAR/Shutterstock.com; p. 29 Peter Hermes Furian/Shutterstock.com.

Printed in China

CPSIA compliance information: Batch #QSENS26: For further information contact Enslow Publishing, at 1-800-398-2504.

CONTENTS

Words in the glossary appear in **bold** the first time they are used in the text.

AT THE ATOMIC LEVEL

Atoms are so small we can't see them with our eyes. But we can see atoms when they combine with many, many other atoms. Atoms usually join to form small **units** called molecules. It's these molecules that, together, are the matter that makes up our world.

LEARN MORE

Atoms are so tiny that we can't see them with most **microscopes**. One hundred million hydrogen atoms lined up are only about 0.4 inch (1 cm) in length!

ATOMS AND ELEMENTS

There are 118 known kinds of atoms. These make up the 118 known elements. Elements are substances, or kinds of matter, that can't be broken down into simpler substances. Gold, carbon, and oxygen are elements. Molecules—combinations of atoms—make up the elements we see around us.

LEARN MORE

Pure elements contain only one kind of atom in their molecules. For example, silver is an element. It only contains silver atoms.

In nature, a few elements are made up of single atoms rather than molecules. Helium is one. In its natural state, it's made up of unattached helium atoms. Some elements are made up of molecules of two atoms bonded together. Other elements bond in larger groups.

DIATOMIC MOLECULES

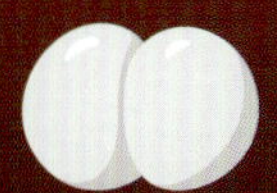

hydrogen (H_2)

nitrogen (N_2)

oxygen (O_2)

fluorine (F_2)

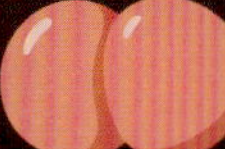

chlorine (Cl_2)

bromine (Br_2)

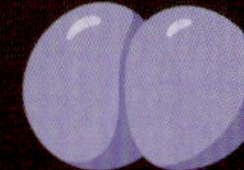

iodine (I_2)

LEARN MORE

A diatomic element is never found as a single atom in nature. It's always bonded to another atom.

Most matter is made of molecules of different kinds of elements. Molecules that contain atoms of more than one element are compounds. When different elements join to become a compound, the elements can change their **properties**. Hydrogen and oxygen, gases at normal temperature and **pressure**, can bond to become a liquid—water!

WATER MOLECULE FORMATION

$$2H_2 + O_2 \longrightarrow 2H_2O$$

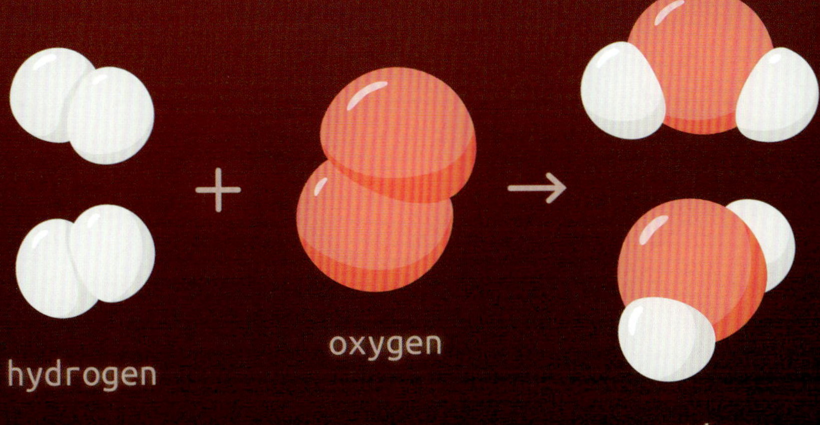

LEARN MORE

When elements mix but don't form **chemical bonds**, they're called mixtures. They keep their properties and can be separated. Air is an example of a mixture of elements.

HOW THEY FORM

So, how do molecules and compounds form? Let's start with atoms. Atoms are made up of smaller **particles** called protons, neutrons, and electrons. Protons and neutrons are found at the center of the atom, called its nucleus. Electrons orbit, or travel around, the nucleus.

ATOMIC STRUCTURE

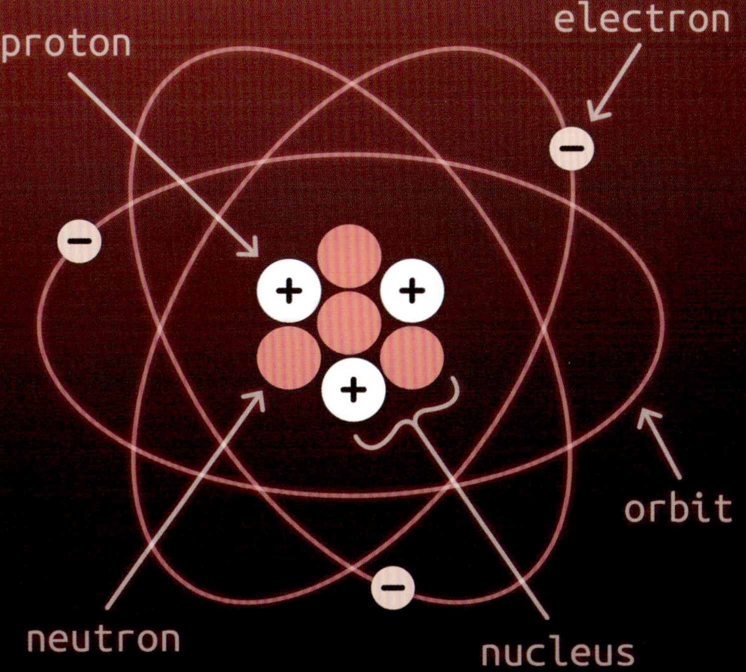

LEARN MORE

Much of an atom is just space. If a hydrogen atom were as large as our planet, the proton in its nucleus would be only 600 feet (183 m) across!

An atom's electrons move around the nucleus in areas called shells. Electrons have a negative electric **charge**, while protons have a positive charge. Neutrons have no electric charge. Opposite charges attract, so the electrons are drawn to the protons. This holds the electrons in place.

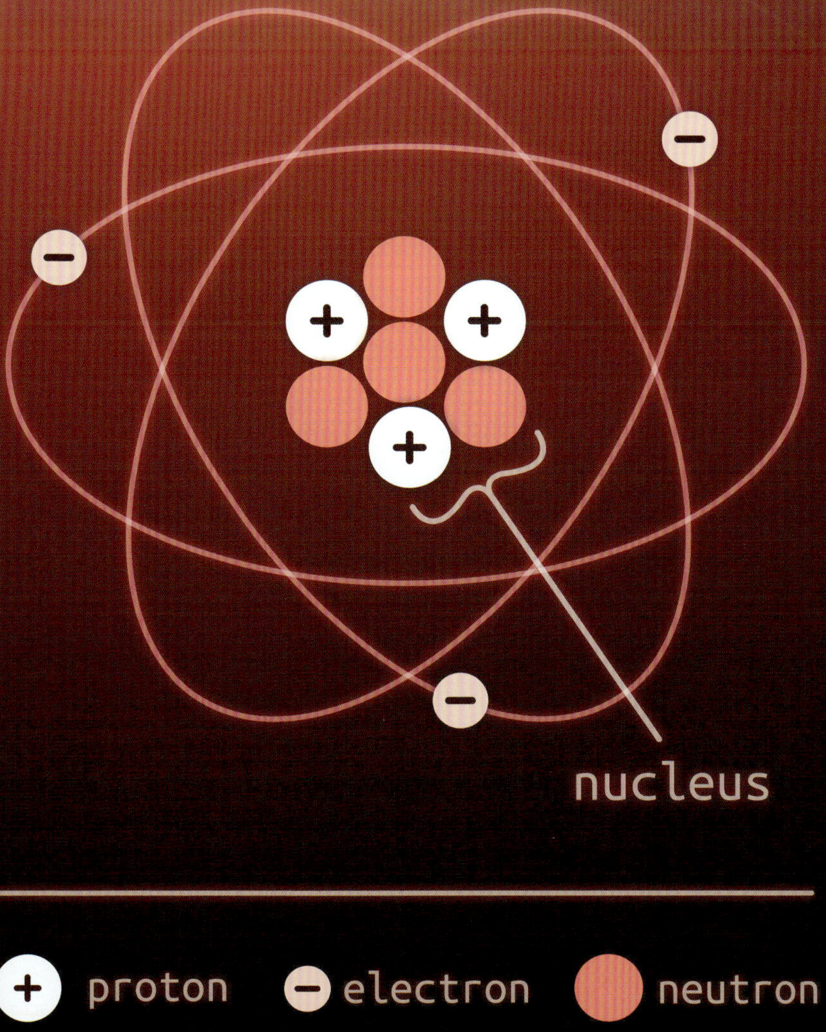

nucleus

LEARN MORE

Protons and neutrons have about the same **mass**, while electrons have much, much less mass.

BONDING

The number of electrons in an atom's outermost shell decides an atom's **reactivity**. An atom can form a bond with another atom by transferring, or moving, an electron to it. This forms an ionic bond. It's how the compound called sodium chloride forms. That's table salt!

SODIUM CHLORIDE

Na + Cl = NaCl

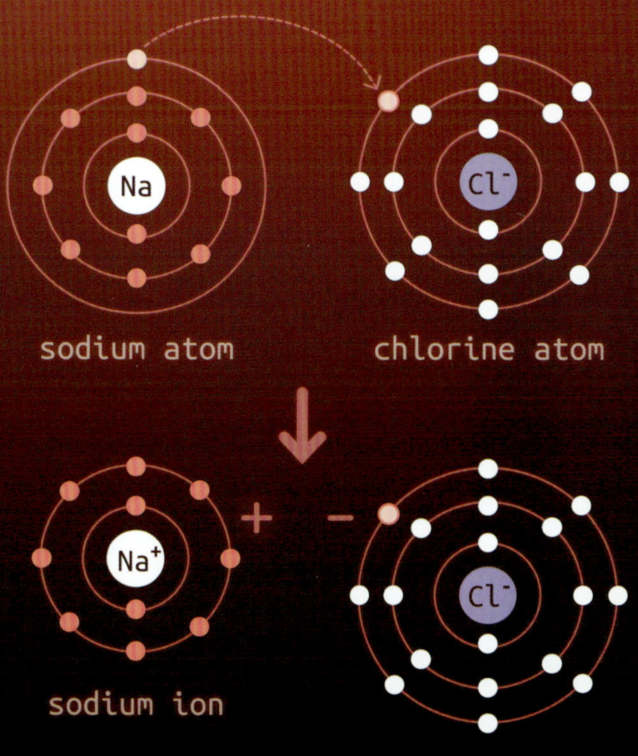

LEARN MORE

If an atom's outermost electron shell is filled, it's less likely to bond with other atoms. For most atoms, eight electrons in this shell make it **stable**.

Sometimes, electrons share two atoms. This forms a strong connection called a covalent bond. Electrons can also be shared among many atoms of metal **ions**. The electrons don't become fixed to one atom but move among them. This is called metallic bonding.

COVALENT BOND

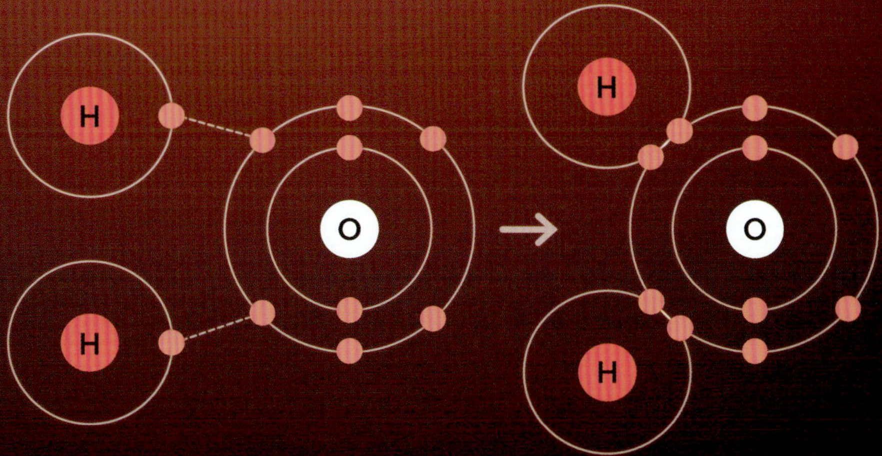

METALLIC BOND

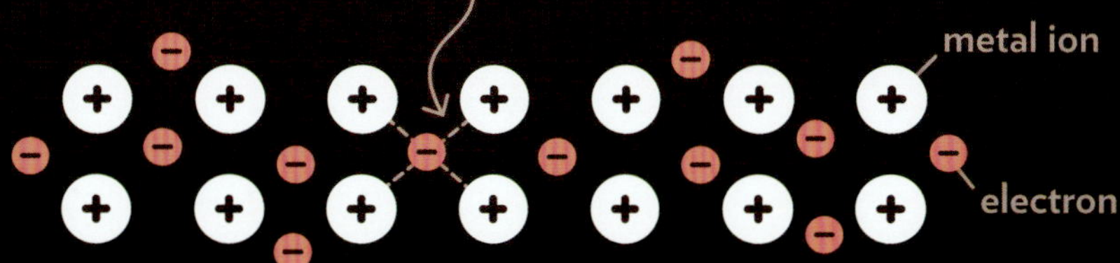

LEARN MORE

Covalent bonds usually form between nonmetal elements. For example, two covalent bonds hold together a water molecule.

MILLIONS OF COMPOUNDS

A compound contains two or more elements but all molecules are **identical**. There are 118 kinds of elements, but many millions of compounds! That's because atoms can combine in so many different ways. Compounds can be made of many different kinds of atoms as well as different numbers of atoms.

LEARN MORE

Carbon dioxide is a compound. Its molecules are made up of one carbon atom bonded to two oxygen atoms. It's a colorless gas in nature. It's an important part of air, but too much of it can lead to problems.

Scientists may call compounds by their chemical **formula**, such as H_2O. This formula tells us the molecule's elements—using their **atomic symbols**—and the number of atoms in each molecule. Some compounds are better known by names rather than a formula, such as the compound called **caffeine**.

THEOBROMINE

$C_7H_8N_4O_2$

carbon
hydrogen
oxygen
nitrogen

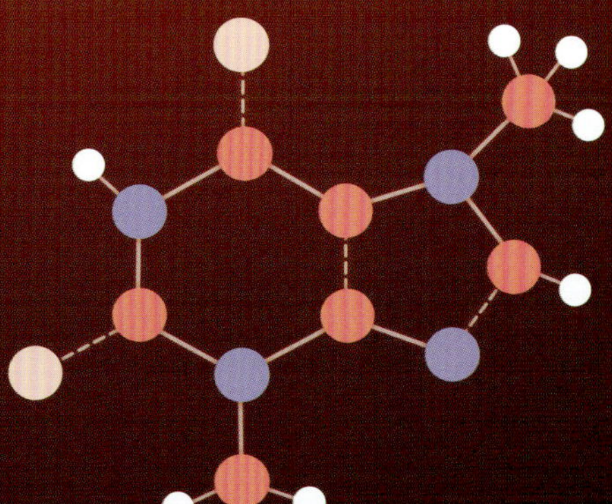

LEARN MORE

Theobromine is a compound made by plants. It's used to make chocolate!

Many, many other compounds are a part of everyday life. But we don't use their chemical formulas, so we don't think of them as molecules. Sucrose is a sugar people bake with. It's a compound of carbon, hydrogen, and oxygen. So without compounds, we wouldn't have cookies!

SUCROSE

LEARN MORE

The chemical formula for sucrose is $C_{12}H_{22}O_{11}$. It has 12 atoms of carbon, 22 atoms of hydrogen, and 11 atoms of oxygen.

ISOMERS

The properties of a compound depend on how the atoms are arranged. The same numbers and kinds of atoms may come together to form different substances. They can have the same chemical formula too. These are called isomers.

LEARN MORE

The sugars glucose and fructose are isomers. Both have the chemical formula $C_6H_{12}O_6$ but have different looking molecules. Both are found naturally in honey and fruits, but fructose is much sweeter.

THE TABLE TELLS US

The periodic table is an arrangement of the elements according to their chemical properties. Where an element is on the table can tell you a lot about its atoms, including how they will react with another kind of atom. Molecules help us make sense of our world!

PERIODIC TABLE

atomic number → 1

symbol → H

name → Hydrogen

1	2	3	4	5	6	7	8	9	10	11	12	13	14	15	16	17	18
1 H Hydrogen																	2 He Helium
3 Li Lithium	4 Be Beryllium											5 B Boron	6 C Carbon	7 N Nitrogen	8 O Oxygen	9 F Fluorine	10 Ne Neon
11 Na Sodium	12 Mg Magnesium											13 Al Aluminium	14 Si Silicon	15 P Phosphorus	16 S Sulfur	17 Cl Chlorine	18 Ar Argon
19 K Potassium	20 Ca Calcium	21 Sc Scandium	22 Ti Titanium	23 V Vanadium	24 Cr Chromium	25 Mn Manganese	26 Fe Iron	27 Co Cobalt	28 Ni Nickel	29 Cu Copper	30 Zn Zinc	31 Ga Gallium	32 Ge Germanium	33 As Arsenic	34 Se Selenium	35 Br Bromine	36 Kr Krypton
37 Rb Rubidium	38 Sr Strontium	39 Y Yttrium	40 Zr Zirconium	41 Nb Niobium	42 Mo Molybdenum	43 Tc Technetium	44 Ru Ruthenium	45 Rh Rhodium	46 Pd Palladium	47 Ag Silver	48 Cd Cadmium	49 In Indium	50 Sn Tin	51 Sb Antimony	52 Te Tellurium	53 I Iodine	54 Xe Xenon
55 Cs Caesium	56 Ba Barium	57 La* Lanthanum	72 Hf Hafnium	73 Ta Tantalum	74 W Tungsten	75 Re Rhenium	76 Os Osmium	77 Ir Iridium	78 Pt Platinum	79 Au Gold	80 Hg Mercury	81 Tl Thallium	82 Pb Lead	83 Bi Bismuth	84 Po Polonium	85 At Astatine	86 Rn Radon
87 Fr Francium	88 Ra Radium	89 Ac** Actinium	104 Rf Rutherfordium	105 Db Dubnium	106 Sg Seaborgium	107 Bh Bohrium	108 Hs Hassium	109 Mt Meitnerium	110 Ds Darmstadtium	111 Rg Roentgenium	112 Cn Copernicium	113 Nh Nihonium	114 Fl Flerovium	115 Mc Moscovium	116 Lv Livermorium	117 Ts Tennessine	118 Og Oganesson

*	58 Ce Cerium	59 Pr Praseodymium	60 Nd Neodymium	61 Pm Promethium	62 Sm Samarium	63 Eu Europium	64 Gd Gadolinium	65 Tb Terbium	66 Dy Dysprosium	67 Ho Holmium	68 Er Erbium	69 Tm Thulium	70 Yb Ytterbium	71 Lu Lutetium
**	90 Th Thorium	91 Pa Protactinium	92 U Uranium	93 Np Neptunium	94 Pu Plutonium	95 Am Americium	96 Cm Curium	97 Bk Berkelium	98 Cf Californium	99 Es Einsteinium	100 Fm Fermium	101 Md Mendelevium	102 No Nobelium	103 Lr Lawrencium

LEARN MORE

All the elements in Group 1 (the first column on the left) have one electron in their outermost shell. They're very reactive. Group 18 have a full outermost shell and aren't very reactive.

CHEMISTRY CONCEPT REVIEW

OXYGEN

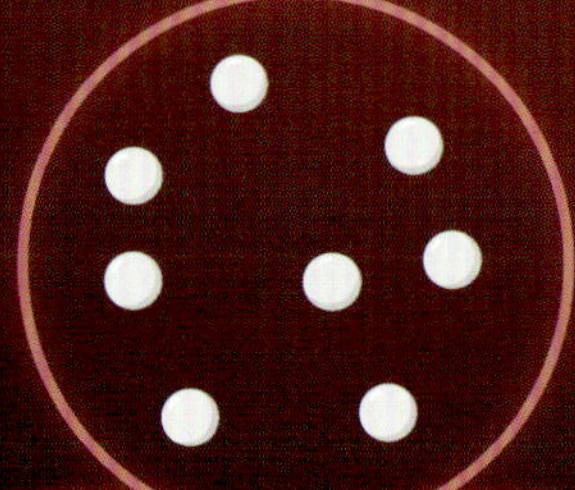

atoms of an element

NITROGEN

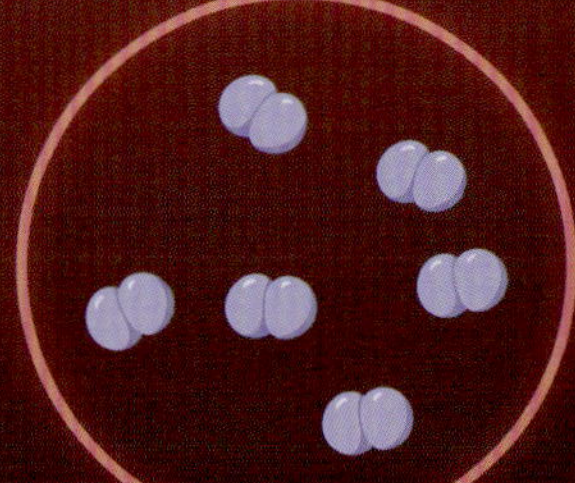

molecules of an element

CARBON DIOXIDE

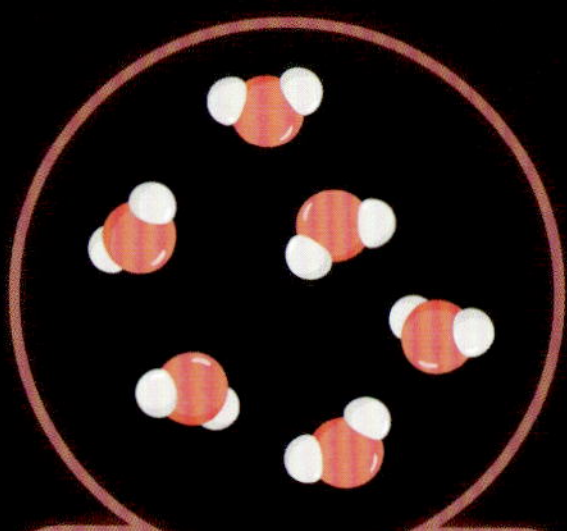

molecules of a compound

AIR

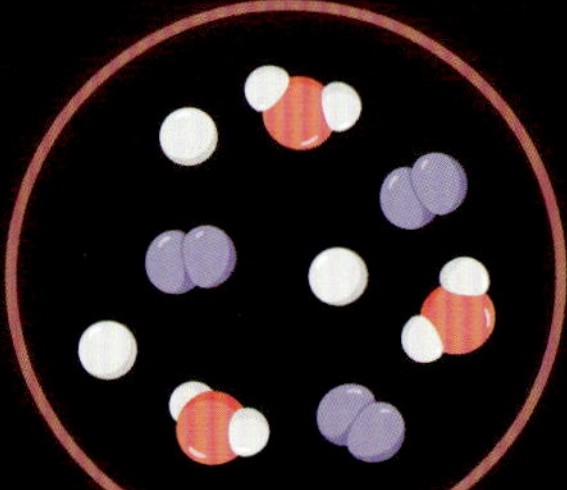

mixture of elements and a compound

GLOSSARY

atomic symbol: One or two letters representing a chemical element.

caffeine: A substance found in coffee and tea that makes you feel more awake.

charge: An amount of electricity.

chemical: Having to do with atomic or molecular changes.

formula: A series of letters, numbers, and symbols showing the chemicals that a compound is made of.

identical: Being the same.

ion: An atom or group of atoms that has a positive or negative electric charge from losing or gaining one or more electrons.

mass: The amount of matter in an object.

microscope: A tool used to view very small objects so they can be seen much larger and more clearly.

particle: A very small piece of something.

pressure: A force that pushes on something else.

property: A special quality or feature of something.

reactivity: How likely a substance is to interact with another substance.

stable: Not likely to change suddenly or greatly.

unit: A group that is part of a larger whole.

FOR MORE INFORMATION

BOOKS

Meyer, Cassie. *Atoms and Molecules.* Chicago, IL: World Book, 2023.

Wells, Robert E. *The Molecules That Make You You.* Chicago, IL: Albert Whitman & Company, 2022.

WEBSITE

Molecules

www.ducksters.com/science/molecules.php

Learn some fun facts, and take a quiz!

INDEX